Swarm Management with Checkerboarding
by J. White, A. Hunt, G. Bannister
ISBN: 978-1-912271-86-3

Published by Northern Bee Books, Scout Bottom Farm, Mytholmroyd,
Hebden Bridge, HX7 5JS (UK). Tel: 01422 882751.
www.northernbeebooks.co.uk
Book design by www.SiPat.co.uk

Swarm Management
with
Checkerboarding

by J. White, A. Hunt, G. Bannister

Diagrams and photographs

Contents

Introduction

For many beekeepers swarming is an annual challenge. For those of us living in close proximity to non-beekeepers, there are particular sensitivities to be observed and swarming keeps us on our toes and preoccupies us through the late spring and early summer. There are whole books dedicated to swarming – the mechanics of it, how to recognise swarming, swarm prevention, swarm control measures and so on.

Our quest, and the reason for experimenting with Checkerboarding, was to try and find a solution to this perennial challenge. We recognise the importance of swarming with respect to procreation and genetic diversity. We also recognise the angst beekeepers suffer when we get that knock at the door from a neighbour that states 'your bees have swarmed into my garden/shed/porch'. And how can we deny that they are our bees?

Even without the proximity of neighbours most beekeepers endure a sustained period of colony inspections, checking for early swarm indicators i.e. queen cells. Therefore, the solution to swarm prevention which does not involve extensive hive manipulations, inspections, mitigation measures, remains one of the holy grails of beekeeping.

Checkerboarding, if it is carried out correctly will prevent swarming and requires no intervention within the brood nest. That is a bold statement. The authors of this document carried out successful Checkerboarding during the December 2019 season on five colonies. Between the three of us we had previously had swarms each year for the last 9 years. Sometimes many swarms per year and sometimes many per hive! In 2020, we had none!

Some of the by-products of Checkerboarding include significant increases in colony size and therefore honey yields. Another is the resident queen is replaced through supersedure.

Checkerboarding is simple to implement. What is more complicated is the understanding of the internal operations of the colony that make this system successful. We have therefore written this booklet with a brief explanation of how to Checkerboard early on. You will also find more details of the colony dynamics and the common terminology used when discussing swarms in the section: 'Notes and Explanations' on page 20.

If Checkerboarding is not for you then there may be a small gem within this document that you could choose to note and be aware of, i.e.: Backfilling, see page 23.

What is Checkerboarding and how do you carry it out?

The essential requirements for Checkerboarding are:

- a super of empty drawn brood frames.

- a super of drawn brood frames with capped honey.

- timing – this must be carried out early, we recommend before February in the UK.

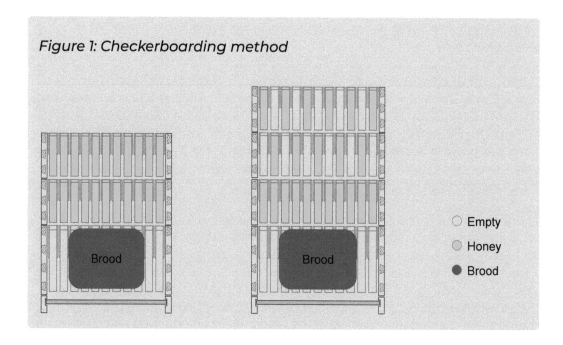

Figure 1: Checkerboarding method

○ Empty
○ Honey
● Brood

The expression 'Checkerboarding' describes the alternating combs that are required for this particular form of swarm prevention. The frames of drawn brood and frames of capped stores are alternated in two supers above the brood nest as shown in Figure 1. Equally this could be two boxes of standard national frames. The alternating of full frames of stores and empty frames is important and is the key to the success of Checkerboarding. This is explained in further detail on page 20. Once a colony has acquired a solid layer of stores above them, typically about one whole super of capped stores, they have created a critical element in the swarm preparation sequence. Our disruption of this layer is a key component in preventing swarming.

Where did this idea originate?

Walter Wright was the first beekeeper to document this method of beekeeping and his writings are freely available on the American bee web site Bee Journal (https://www.beesource.com/threads/walt-wright.365657/). He took up beekeeping in his late fifties having retired from NASA where he was a troubleshooting engineer. Wright canvassed beekeepers and asked them what their main issue with beekeeping was. Swarming featured predominantly. He then turned his investigative skills to this issue and came up with the concept of Checkerboarding.

He declined the advice of fellow beekeepers, books and mentors and preferred instead to approach beekeeping using his observational skills without any preconceptions. Within 3 years he was convinced he had resolved the conundrum of swarming. By resolved we mean that he had enough understanding of the primary drivers and mechanisms for swarming that helped him come to a low intervention resolution with absolute success i.e. Checkerboarding.

So evident did the understanding of honeybee behaviour and the solution to swarming seem to Walter Wright, that he could not fully appreciate why it had taken thousands of us beekeepers looking into millions of hives so long and still without realising what he thought was obvious. We think it is safe to say that Walter Wright was a self-assured person. He concluded that beekeepers "see but they do not observe, or ask themselves why the bees do what bees do". The authors of this booklet, based on our own beekeeping, would concur with this. We believe that he was successful in his conclusions and subsequent practices.

Walter Wright believed that honeybees are motivated by survival first. In early spring, long before the peak swarm period, some colonies will have already committed to their second priority: generating swarms for reproduction. He concentrated his investigations on the colony activities that support these two colony goals.

We believe that his thinking and conclusions represent a significantly different approach to understanding honeybee behaviour compared with conventional swarm management advice and practices. Amongst other things he advocated that nectar in the brood nest (backfilling) and bee crowding, which result in "congestion", are essential steps towards swarm production. Wright went on to distinguish between congestion in the brood nest and overcrowding (too many bees in the colony), which can cause swarming, typically later in the year.

Checkerboard examples:

The mechanics of Checkerboarding are very simple. Figures 2 and 3 below show two scenarios to illustrate this. It is not important whether you begin with an empty frame or a frame full of stores, what is important is the frames are alternated.

If you are Checkerboarding and applying fondant to your colony as part of winter provisions, then please do not place your crown board immediately above the brood with the fondant on top of the crown

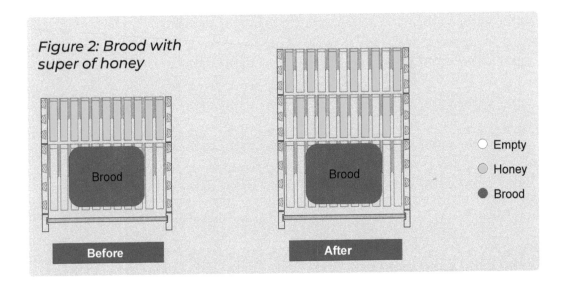

Figure 2: Brood with super of honey

Before

After

Empty

Honey

Brood

board. Doing this will prohibit the movement of the bees and their perception of the overhead Checkerboarded frames. Instead place your fondant directly onto the top bars, within an eke, and put your crownboard above your topmost Checkerboarded box (Figure 3).

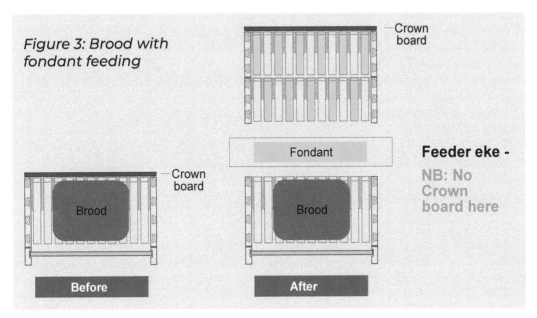

Figure 3: Brood with fondant feeding

Figure 4 shows how Checkerboarding might be achieved using a super of stores and one super of drawn comb.

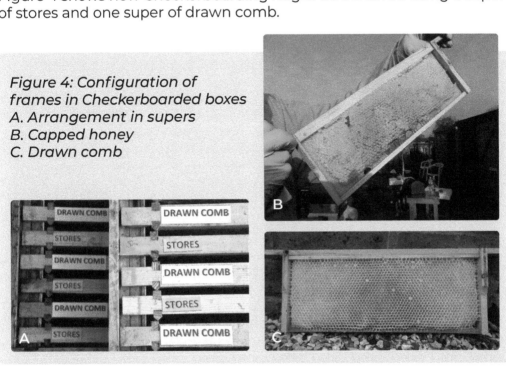

Figure 4: Configuration of frames in Checkerboarded boxes
A. Arrangement in supers
B. Capped honey
C. Drawn comb

The Checkerboarding year

To clarify this process, we have divided up the year into key Checkerboarding milestones to help illustrate what needs to be done and when. We have deliberately kept this brief. More detailed explanations can be found on page 20 in 'Notes and explanations'.

1. Late summer/autumn – Build up your brood frame reserves

2. December – Carry out your Checkerboarding

3. Spring Management – retain the empty brood frames above the expanding colony

4. Summer - Nectar management

A more detailed timeline (Figure 5) showing the Checkerboarding year is available on pages 16 and 17.

1. Late summer/autumn - Brood frame reserves

This requires some forward planning. In order to carry out Checkerboarding a reserve of frames is required. This Checkerboarded structure achieves two objectives: very early on it breaks up the reserve of honey that the bees perceive as an essential requirement before they can swarm; and, it provides the framework for the brood to rapidly expand during late winter/early spring. A shallow box (or standard national or equivalent) of empty drawn brood and another shallow box of drawn brood frames with stores are needed for each hive by the time that you wish to Checkerboard. If it is your intention to Checkerboard it is worth giving this some thought as the summer comes to an end and you begin your autumn feeding programme. The aim here is to provide a reserve of food within brood frames that would not normally be used (by the beekeeper) to store honey. This honey will provide valuable food and once displaced will be populated with brood.

2. December - Checkerboarding

The key to success is in the timing. The process of Checkerboarding has been outlined above. The act of Checkerboarding needs to be carried out at least 9 weeks prior to peak swarm season, which Walter Wright identified as peak apple blossom period (mid-season flowering apples). Owing to the vagaries of our seasons, we removed this uncertainty and carried out Checkerboarding before Christmas i.e., any time during

Figure 6: Checkerboarding requirements

December, and if possible, when temperatures are not too cold as illustrated in Figure 7.

This figure shows the timeline and the key milestones during the swarm process. At a push this deadline (December) could be extended into January, but with cooler temperatures and seasonal variations in swarm timings this should only be a fallback. See more in: 'Notes and Explanation, Point a.' on page 20.

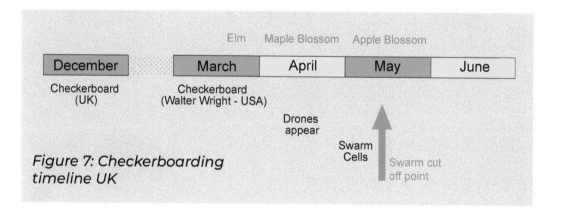

Figure 7: Checkerboarding timeline UK

3. Spring Management

Having Checkerboarded the hives before Christmas the colonies can then be left alone until spring when rapid colony expansion will occur. Once the nectar flow in the spring begins, it is important to ensure that there are two empty supers with drawn comb above the growing bee colony. To be clear, as soon as the bees allow the queen to lay and occupy the lower of the two Checkerboarded supers, then add another super of drawn comb/foundation at the top. The hive on the right of Figure 8 shows the rapid spring expansion. We have left the original Checkerboarded frames in this picture. In reality this colony would have nectar, honey and pollen around the brood similar to Figure 18: Drone mini cluster.

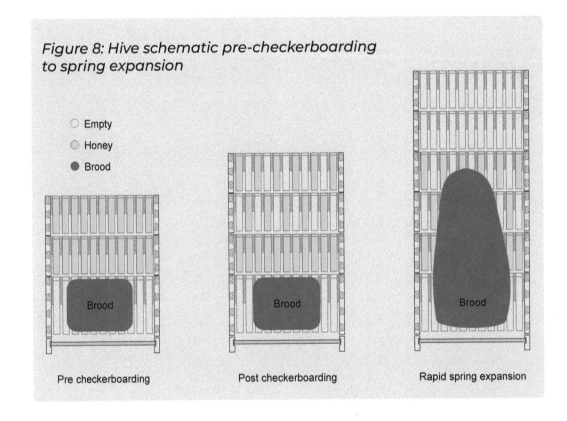

Figure 8: Hive schematic pre-checkerboarding to spring expansion

○ Empty

○ Honey

● Brood

Brood

Brood

Brood

Pre checkerboarding   Post checkerboarding   Rapid spring expansion

3.1 Drone provision

Drones are worth a special mention.

Figure 9: Drone close up

We can help the colony by placing sufficient drone comb within the brood area. We recommend providing a single frame of drone comb on the outside edges of each brood box during the spring. After populating these frames with drones, these frames will provide a useful nectar reserve going into winter and then in the spring these frames can be freed up for drone laying when more drones are required.

There is also another advantage in using the outside frame for drone management. It is easy to remove these frames should you choose to sacrifice some of your drone frames as part of your strategy for managing varroa mites..

Figure 10: Drone provision

4. Summer – Nectar management

During the main flow, continually monitoring and adding supers is essential in order to prevent the colony becoming overcrowded and also preventing the upper most boxes being filled with capped honey and the bees 'thinking' they have reached the top of their colony. Both of these events could trigger swarming. At this point we are concerned with managing the flow of nectar that is incoming.

Figure 11 shows the colony expanding into the previously Checkerboarded frames with two empty supers at the top of the colony. This additional expansion area allows plenty of spare capacity for the bees to collect nectar in and will prevent swarming.

During 2020 we used step ladders to reach the upper boxes of our colonies.

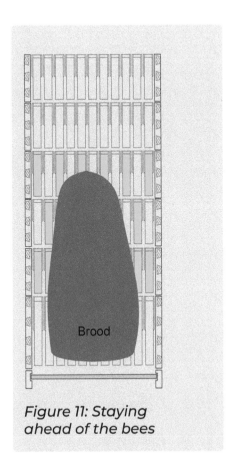

Figure 11: Staying ahead of the bees

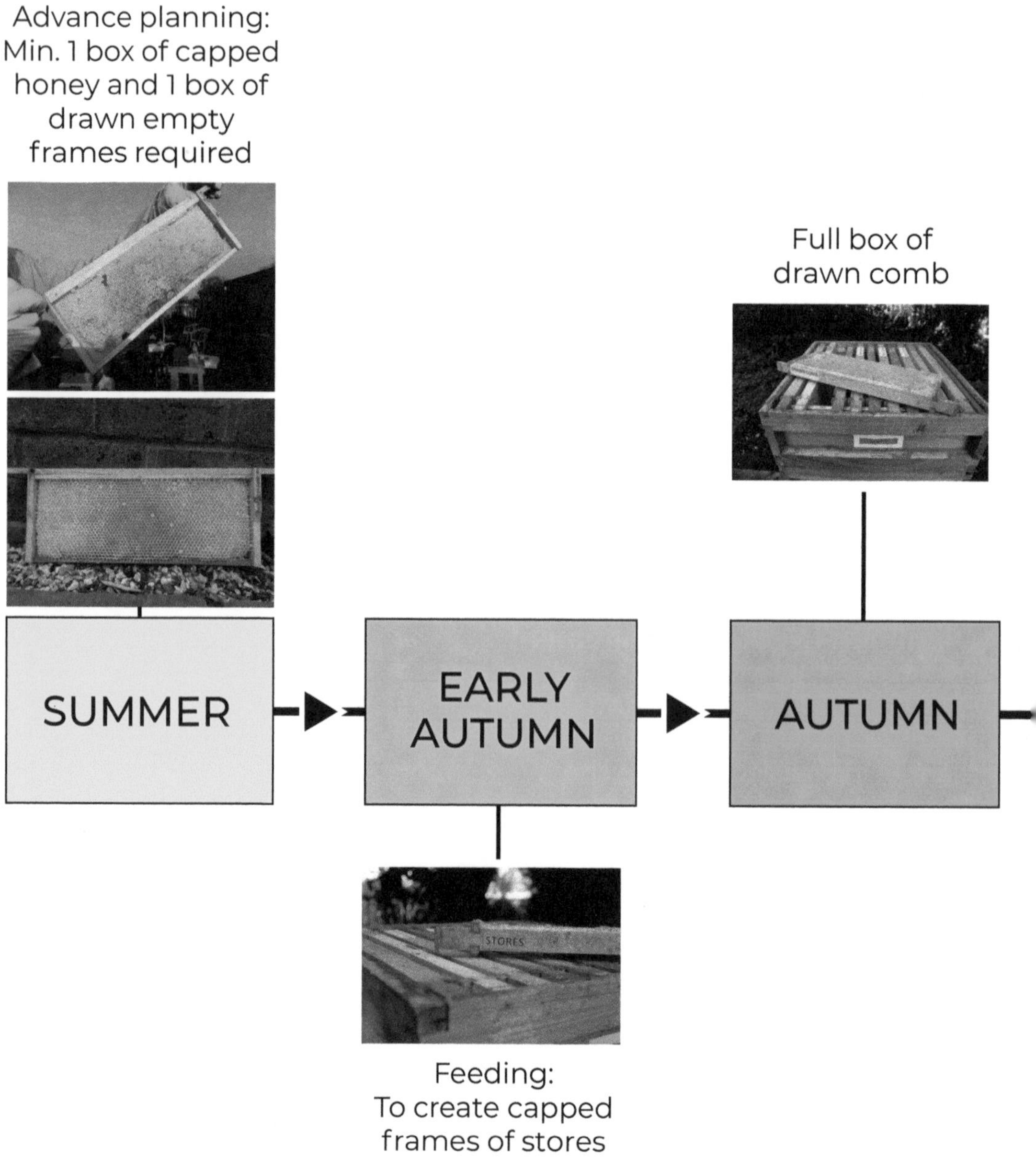

Advance planning:
Min. 1 box of capped
honey and 1 box of
drawn empty
frames required

Full box of
drawn comb

SUMMER ▶ EARLY AUTUMN ▶ AUTUMN

Feeding:
To create capped
frames of stores

Figure 5:

The Checkerboarding year at a glance

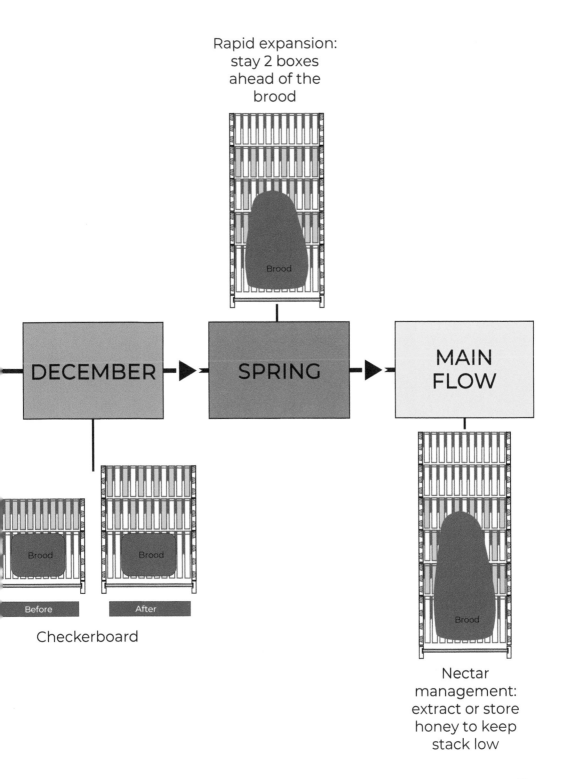

Rapid expansion: stay 2 boxes ahead of the brood

Brood

DECEMBER

SPRING

MAIN FLOW

Brood

Brood

Before

After

Checkerboard

Brood

Nectar management: extract or store honey to keep stack low

We allowed our hives to get to reach tall heights of up to 11 or 12 supers, see Figure 12.

Figure 12: The beehives at their tallest

The high colonies were a result of following the original methodology and letting them grow, which they did......mightily! We quickly concluded that for safety reasons and in the interest of retaining a sound back, shorter colonies would be a much more sensible way forward. Managing heavy supers on a stand or step ladder was not prudent or safe.

We began to appreciate why Walter Wright changed his terminology from Checkerboarding to Nectar Management. Once we get to the main flow that is exactly what we are doing, managing the incoming nectar and subsequent honey production.

Our strategy for coping with this increasing honey crop was two-fold: 1. Periodically removing some of the filled supers and extracting the capped honey; or, 2. Removing ripened honey and storing this on non-Checkerboarded colonies if you have them available. Doing this greatly reduces the hive 'tower' down to five or six supers above the brood. Typically, we would take off the two lower capped honey supers and then put two more supers on top either immediately (if you have spare boxes), or extract those supers and return them the next day.

End of Year Colony Size

Despite the tall colonies that have been generated during the spring/ summer, by the time we get to winter the colonies have shrunk right back to a 'regular' winter size. At the time of checkerboarding in December 2020 the colonies were typically the size as illustrated in Figure 13.

Figure 13: The beehives in winter

Notes and explanations

a. December - Checkerboarding

One of the advantages of Checkerboarding is that it does not involve disturbing the brood area at all. Checkerboarding is carried out above the brood nest within the honey reserve. Walter Wright understood that one of the key triggers to swarming occurs once the brood nest has reached the capped honey above. Initially he experimented by doing what many of us do, increasing available brood space in order to attempt to offset swarming. He did this initially by using two brood boxes and then tried a third brood box to increase available space further. The bees still swarmed, albeit less so. Clearly having more available laying space for the queen does have an impact on swarming but importantly it is not the key factor.

In the majority of cases the presence of a solid layer of available stores, at the top of the colony, and the bees awareness of this layer, is the critical factor which induces the bees to begin swarm preparations. Disrupting this layer by Checkerboarding is part of the solution but remember it takes time for the bees perception of the top of the colony to change which is why Checkerboarding must be done well in advance of the swarm season.

The Checkerboarding manipulation raises the apparent top of the capped honey reserve, above the brood nest, and spreads it across two boxes in alternating frames of stores and empty drawn frames. It is this disruption of the solid stores that inhibits swarm preparation providing it is done well before the swarm season. If we left this solid capped reserve in place it is possible that the bees would begin swarm preparations.

b. Spring management

As the colony population increases, for a given brood volume the cluster of bees is approximately one and a half times this volume. These are the bees that would generate overcrowding had you not provided additional space. The colony is now storing nectar above the brood nest and this serves two separate purposes: firstly, it prevents the tendency to begin swarm preparations because they have this space available to fill with nectar (see backfilling on page 23); and secondly, it provides frames of nectar to continue brood nest growth.

At this point in the year continuous adding of boxes is essential in order to prevent the colony swarming. The growing population along with available stores and overall cavity space has implications for adding boxes. If you wait until the last super is nearly full before adding another box, the bees will be reducing their brood size (backfilling) and swarm preparations may well begin. During a good flow (spring or summer) this could be as frequent as a single super each week. If in doubt add an extra box.

As the cluster volume increases, more and more nectar is accumulated above the brood nest and underneath the clustered bees. One of the features of Checkerboarding is the accumulation of this nectar during the build-up. More bees usually result in more honey. The larger the brood nest, the more nectar is in place prior to the main flow and this is what is happening here. The collection of nectar during this period occurs when most traditional swarm prevention techniques would have us reducing colony size or disrupting the brood and slowing down the bees development.

For colonies that are planning to swarm, they expand the brood nest to the honey reserve (e.g., this may be a fully capped super). By this time, they will have increased population to a safe limit (i.e., enough bees to swarm) and are now able to move into the swarm preparation phase. However, they are not ready to start producing swarm queen cells yet. Imagine what would happen if approximately half the colony left at that point! The bees left by the departing swarm would be hard pressed to support the original brood volume due to a shortage of nurse bees and foragers. There would not be sufficient population to maintain brood nest temperature for that volume of bees and brood. Therefore, brood nest reduction is required (backfilling) and this is a key indicator that the colony is beginning swarm preparations.

It is important to maintain empty boxes of drawn comb above the cluster, if possible, until the main flow. After this time, foundation can be used. Remember, as long as the colony is expanding the brood volume, it will not generate a reproductive swarm and, by deliberately maintaining storage space at the top of the brood, overcrowding swarms are avoided (see Figure 14).

To check the development of the brood nest through the swarming season, the beekeeper only has to penetrate the hive to the top of the brood. The colony expands the brood nest in small discrete steps at the top. Remove the accumulated nectar down to the top of the brood. If you lift out a frame with the arc of the brood expansion dome defined, the top of the brood area becomes obvious.

Figure 15 shows the top of the brood and a band of cells referred to as 'the drying cells'. These cells provide a band for the brood to expand into. Brood is below this expansion band and full cells of nectar above. The expansion band can range through reduced nectar, drying

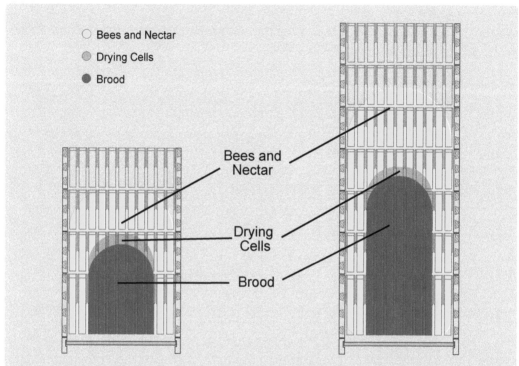

Figure 14: Colony expansion. Cluster development during build-up

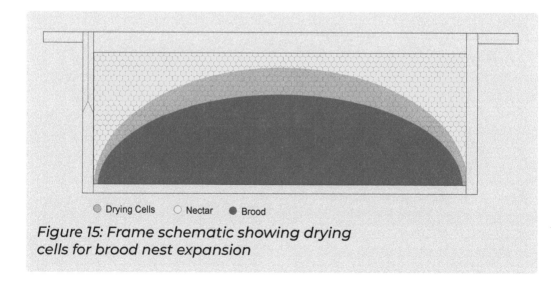

Figure 15: Frame schematic showing drying cells for brood nest expansion

cells, eggs laid, or open larval brood, depending on the stage when inspected. At any of the stages of expansion, the band is obvious at a glance. Walter Wright described this as "drying cells for expansion." The expansion band (with cells drying out) at the top of the brood nest is evidence that the colony has no intention to swarm. Drying cells above the brood is exactly the opposite of nectar congestion (backfilling) in the top of the brood nest.

c. Backfilling – an early swarming indicator

One of Walter Wright's many gems is the concept of backfilling. We include this as an early swarm indicator. Even if Checkerboarding is not for you this may well be worth a note.

For a colony that is going to swarm, once the colony that is intent on swarming has fully expanded, they cannot swarm yet. There is an in-built requirement to preserve the life of the parent colony. They do this by reducing down the volume of the brood nest so that once the new queen begins laying the number of available cells is reduced by about 50%. Figure 16 illustrates this. You can observe the filling in of cells with nectar in the centre and top left of the frame.

This reduction is essential if the remaining bees are to support newly laid eggs. Therefore, the first noticeable sign of swarm preparation is this reduction in the brood volume by filling with nectar or pollen. Walter Wright referred to this as 'backfilling'. This tends to happen from the top of the brood nest and working down.

Backfilling therefore achieves a number of objectives:

- Reduction in the available brood nest to cater for the reduced colony size.

- It provides a reservoir of liquid feed for the bees that are about to depart with the swarm, especially the wax makers.

- Reduced area for the queen that is about to leave to lay in.

At no time has the parent colony survival been jeopardised. The swarming process has evolved to ensure this. Backfilling is not restricted to the period of brood nest reduction associated with swarming. Anytime the brood volume is reduced backfilling is applied. It is also applied in autumn as the brood nest declines.

Figure 16: Backfilling with nectar on a brood frame

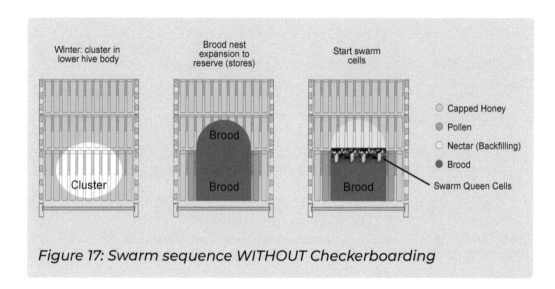

Figure 17: Swarm sequence WITHOUT Checkerboarding

Therefore, the first noticeable sign of swarm preparation is this reduction in the brood volume by filling with nectar or pollen.

This serves as an early indicator that the colony is making preparations to swarm and is well worth looking for if you want to prevent swarming early on in the process.

d. Queen excluders

We tend not to use these as it can constrain the brood nest.

Walter Wright, in his writings, considered queen excluders to be an 'insidious devices' and did not like them or use them. He considered a prominent fault of excluder design is that there is no relationship between comb spacing and excluder wire spacing. Excluder wire spacing has no regard for openings in comb spacing above and below. They generate additional lateral travel for the bees above and below the excluders and more space is required to avoid congestion. One of his tips was to consider adding space on the upper side for those with top space hives. Below the queen excluder the bees manage by adding some brace comb but above with limited space there is lateral movement and congestion. Adding space beyond the 'bee space' is something of an anathema for many of us beekeepers but he may well have a point. In his opinion queen exclusion should have been the starting point in device design, and not the only consideration.

As our bees expand in spring many of us add a queen excluder and super. This often acts as a lag in the bees getting started through the excluder. The bees are reluctant to immediately traverse the excluder and begin storing nectar above.

In Walter Wright's writings he stated: "for excluder addicts, there is a simple way to start traffic through the device. You cannot hide brood or honey above the excluder. Raise a frame of brood or honey above it at installation, and traffic through it starts promptly". Many beekeepers already recognize this trick within their husbandry.

Once the main honey flow significantly begins, and the brood nest has stopped expanding, a queen excluder can be inserted. At this point, with a flow on and the brood nest fully expanded, the bees are unlikely to take the queen above the honey barrier and lay in cells above. In particular if you have not prevented them from rearing a comfortable number of drones in the basic brood chamber, there will be no drone brood in the supers.

e. Drones

Increasingly we are more aware as beekeepers of the importance of drones and their value in terms of individual colonies and bees more widely.

With the approaching swarm season in spring, colonies begin drone production with some urgency. During most of the active brood rearing season drones naturally make up 15% to 20% of brood comb, if not more. If provision for drone cells is too low the colony can improvise and not without consequences to the beekeeper! Sometimes bees will use the inter-frame spaces for drone rearing i.e.: on the top bars and also converting regular brood frame cells to drone rearing.

Within feral colonies drones are located at the periphery of the comb and sometimes in 'satellites' away from the brood nest in a hollow within a branching limb.

If within our hives bees find drone size cells in the overhead capped honey (some use these for honey production foundation), and we the beekeepers have not made adequate provision for drone laying, they will use these cells for the queen to lay in. Even when these cells are outside the cluster and filled with nectar or honey, they will move or consume these stores and prepare those cells for eggs. They do this with a remote mini cluster and escort the queen up there to lay. A contingent of bees maintains the mini cluster to care for and warm the drone brood. This is exactly what happened when we did not provide sufficient drone comb during the period of rapid cluster expansion in mid-April 2020 (Figure 18). Here you can clearly see the satellite of drone cells above the honey barrier. We can avoid this and help the colony by placing sufficient drone comb within the brood area.

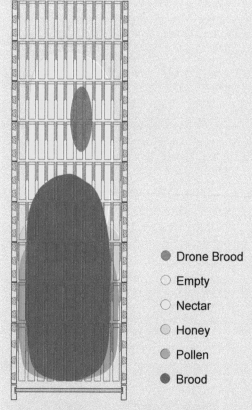

- ● Drone Brood
- ○ Empty
- ○ Nectar
- ○ Honey
- ● Pollen
- ● Brood

Figure 18: Drone mini cluster

f. The honey reserve

The bees are able to monitor the volume of available honey within a colony, and they maintain a reserve amount. Walter Wright found that the bees reserve is equivalent to a shallow super's worth of honey. This is their emergency food store, and they are very reluctant to go into this reserve. A large brood rearing colony can consume a quarter of this reserve in a day. So, 'the reserve' is a minimal amount required in order for normal colony operations to continue, especially when conditions become difficult.

This reserve honey is an emergency food source. When the bees have to tap into these reserves, brood nest expansion stops. If the reserves are further depleted, brood rearing stops. Beyond that, brood cannibalism and hive malnutrition rapidly occur. Once a hive dips into its reserves, the colony's maximum seasonal potential is significantly reduced.

g. Swarm prevention techniques

These are designed to prevent swarming and they are many and varied. Examples include adding supers to a colony and any deliberate disturbance of the brood nest e.g., removing frames of brood and donating them to other colonies or adding additional brood capacity.

Other more drastic brood nest disruption techniques reduce the number of bees available or slow down the expansion of the nest. Examples of these manipulations range from brood box reversal to more invasive measures such as a Demaree manipulation. All of these manipulations serve to reduce the number of bees (or slow down their expansion). In doing so this also reduces honey production.

h. Congestion

For the purpose of this document, we define congestion as being:

1. Lots of adult bees within the colony

2. Nectar in the brood area

This congestion is the honeybee's natural state. It is worth noting that bees produce more honey when they are 'congested'. Congestion is different to overcrowding which can lead to 'overcrowding swarming'. Congestion is NOT the cause of reproductive swarming.

Neither form of congestion causes reproductive swarming. The build-up process for swarming does not create an intolerable level of bees, it is in fact a controlled process. The first activity of swarm preparations that we the beekeepers may notice is backfilling (see page 23).

i. Reproductive swarms

This is a natural function for honeybees and their main form of procreation. It usually occurs early in the season.

After colony survival reproductive swarms are usually the primary motivation of the over wintered colony. Generating such a swarm will not jeopardise survival of the existing colony. If for any reason swarm preparation plans are aborted the bees can always try again the following year. Only the colony that can afford it, in any given year, will produce a reproductive swarm and to do so they need a predetermined amount of stores and bees in order to swarm safely. Checkerboarding disrupts this process early in the swarm preparation cycle.

j. Overcrowding swarms

These usually occur later in the season and are motivated by colony survival.

These swarms are effectively expendable if they have been issued to alleviate overcrowding once the colony population is out of proportion to the available stores and space. Here the colony is thinking solely about the parent colony survival as a priority. For such colonies, issuing a swarm will help to ensure this by taking the pressure off stores and space. These swarms are effectively sacrificial.

As with all things there are exceptions. For the majority of honeybees, we would suggest the above processes normally apply.

Finally

Walter Wright was, in our opinion, a truly great beekeeper and very little of what he recorded has been absorbed and practiced by beekeepers. The credit for this booklet is really down to him and his amazing observations and recording. We will finish with a quote from Walter Wright:

"The fact that you will find none of this information in your favourite reference book does not make it any less true. If you look for the effects in your hive, you can confirm that the concepts are valid".

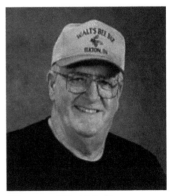

Figure 19: Walter Wright. Photograph taken from the American bee web site Bee Journal: www.beesource. com/threads/ walt-wright.365657/

Glossary

Backfilling: This happens when the brood area is reduced by the bees filling it with nectar. It is not restricted to the period of brood nest reduction associated with swarming but can happen at any time the brood volume is reduced i.e.: in the autumn as the brood nest declines.

Brood nest: This is the area in the hive where the Queen lays her eggs.

Checkerboarding: Is a method of swarm prevention. The mechanics are quite simple. It is when a Super of empty drawn brood frames and a second Super of drawn brood frames with stores, are alternated in the two Supers and placed above the brood nest. This can also be achieved using any size frames.

Drawn comb: This is a frame that the bees have already used and has been cleaned for re-use.

Drone comb: Specific foundation that has been drawn by the bees with large cells to accommodate the laying of Drones.

Foundation: This is clean wax sheeting used to make up new frames.

Nectar management: The process of managing the incoming nectar and subsequent honey production by adding Supers and extracting honey at the appropriate time.

Queen excluder: A grid used to keep the Queen in the brood area and separate from the honey Supers.

Swarming: This is the honeybee's natural survival instinct and occurs when the old queen leaves the hive with some of the bees. Initially they will find somewhere to hang in a cluster while the scout bees find a permanent new home. A honeybee colony swarming is a natural process and can be caused by a number of reasons.

Acknowledgements

Our sincere thanks go to James Warren of Cambridge Beekeeping Association for his editing advice; to Stuart Hunt for interpreting our ideas and designing the diagrams; and, to Nicola Bannister for her help with the design of the front and back covers as well as taking the photograph on the back cover.

Printed in the USA
CPSIA information can be obtained
at www.ICGtesting.com
LVHW081737271223
767386LV00029B/1882